31 THINGS TO KNOW ABOUT TIME

By

Nicolas Comyn

Copyright © 2024 Nicolas Comyn

All rights reserved. No part of this book may be reproduced, distributed, or transmitted in any form or by any means, including photocopying, recording, or other electronic or mechanical methods, without the prior written permission of the author, except in the case of brief quotations embodied in critical reviews and certain other non-commercial uses permitted by copyright law. For permission requests, please contact the author through info@40thingstoknowwhenyour40.com

Published by Nicolas Comyn

Ireland

Disclaimer: This book is intended to provide helpful and informative material on the subjects addressed. It is not intended as a substitute for professional advice. The author and publisher specifically disclaim any liability that may be incurred from the use or application of the contents of this book.

This book is dedicated to:

My Wife Hazel, who has always made time an absolutely wonderful experience.

Contents

Page

- 5. Time Dilation – Why the Future is Closer to the Floor
- 8. The Cosmic Calendar – You're Fashionably Late
- 10. The Twin Paradox – Science's way of Saying I'm Younger than You
- 12. Now Only Lasts 3 seconds – The Ghost of Present Past
- 14. Time Isn't What it Seems – The Elasticity of Now
- 17. Time as a Commodity – The Great Internet Heist
- 19. Before Time Zones – The Wild West of Clocks
- 22. Time and Navigation – The Longitude Lottery
- 26. Ancient Sundials – When Time was Literally "Made in the Shade"
- 28. The Secret Lives of Body Clocks - Humans, Animals, and the "Time Zone Zoo"
- 31. Time and the Divine - When Eternity Is Just the Beginning
- 34. Time Etiquette Across Cultures - The Art of Being Fashionably Late (or Early!)
- 37. Is Time Even Real? Or Are We Just Really Good at Pretending?
- 40. Timekeepers Through the Ages - From Burning Ropes to Atomic Clocks

43. Time Travel - The Myths, Legends, and the Bizarre Believers

47. Dreamtime - Where Minutes Become Hours and Logic Takes a Nap

50. Time Loops and Déjà Vu - When Life Feels Like a Rerun

53. Chasing Immortality - The Timeless Quest to Outsmart Aging

56. Talking Time - When Languages Bend the Clock

65. Memory - The Time Machine in Your Mind

73. Time and Rhythm - When Music Messes with the Clock

80. Biological Clocks - Nature's Strange Timetables

83. The Speed of Light - When Time Stands Still (Sort Of)

90. Aging in Time - The Strange Clock Inside Us

94. Lost in Space (and Underwater)- How Extreme Environments Warp Time

97. The "Butterfly Effect" and Time - How Small Actions Can Change Everything

101. The Psychology of Time

104. Why Does Time Fly When You're Having Fun?

107. The Time of Your Life - How Moments Make Memories

110. Time Capsules - Messages to the Future

114. Time in the Movies - Hollywood's Take on Time Travel and Beyond

Introduction

"Time Waits for No One, or Does It?"

Time. It's the tick-tock of the clock on your wall, the swirl of the stars above, and the subtle stretch of moments when the Wi-Fi goes down. It's an enigma we measure, treasure, and often misuse—sometimes all at once. But what is time, really?

This book isn't here to give you a dry lecture on seconds, minutes, and hours. Instead, it's an exploration—a playful, sometimes mind-bending journey into how time shapes our universe, our cultures, and even our very selves. From the dizzying realms of Einstein's theories to the quirky rhythms of your body clock, we'll uncover the secrets of a dimension that seems so familiar yet remains utterly mysterious.

Expect to laugh, ponder, and perhaps question everything you thought you knew about time. We'll traverse cosmic calendars, hop through time loops, and even visit the Wild West of clocks before time zones civilized chaos. Each chapter is a step through the looking glass, where science meets wonder and history meets hilarity.

So, whether you're a seasoned time enthusiast or just someone wondering why Mondays last forever while weekends vanish, welcome aboard. Strap in—it's going to be a timeless adventure.

Time Dilation
Why the Future is
Closer on the Floor

Ever wondered if you could visit the future just by lying down? In a way, you can, thanks to time dilation, a mind-bending concept straight out of Einstein's theories.

Here's the deal: time runs more slowly the closer you are to a massive object, like Earth. This effect is so real that even the difference in height between your head and your feet creates a tiny bit of time travel. If you're lying on the ground, time is moving slower for you than if you were standing up. The difference is microscopic, but technically, a part of you is living in the future every time you stretch out on the couch!

Now, scale this up to cosmic levels. Near the intense gravity of a black hole, a clock would tick far slower than one on Earth. So slow, in fact, that while you spend a few hours there, decades pass here. Time becomes a twisted, almost absurd notion: the closer you get to the black hole, the more you watch the rest of the universe's history flash by, like an intergalactic fast-forward button.

So next time someone says they're killing time by lying down, remind them they're actually leaping ever-so-slightly ahead into the future!

The Cosmic Calendar

You're Fashionably Late

Imagine if the entire history of the universe were compressed into a single calendar year, from January 1st to December 31st. It's the "Cosmic Calendar," and trust me—it makes you feel like we humans only just showed up to the party.

On this scale, the Big Bang happens at midnight, January 1st. Stars are born in early January, galaxies form by spring, and our solar system finally shows up in early September. Dinosaurs don't even appear until December 25th (yes, they get the holidays!), and humans? We strut in just 23 seconds before midnight on New Year's Eve.

So if you think you're fashionably late to most things, take comfort—our whole species basically showed up to the universe's party at the last possible moment. Even stranger? Recorded human history—the entirety of everything written down—is just the last second of the cosmic year.

> When you think about it, we're kind of the ultimate last-minute guests in the cosmic timeline.

The Twin Paradox
Science's Way of Saying,
"I'm Younger Than You!"

Imagine this: you and your twin are celebrating your 21st birthdays together when one of you decides to go on an interstellar joyride, traveling near the speed of light for a few years. When they return to Earth, here's the twist— they're younger than you.

It's called the "Twin Paradox," and it's a trippy result of Einstein's relativity. Time actually slows down when you're traveling fast enough. So, while Earth-twin ages as usual, Space-twin barely ages at all. This cosmic loophole means the next time they have a birthday, they might need an "Under 25" ID to rent a car while Earth-twin's already on their way to 30!

In a strange way, relativity is like the universe's way of saying, "Yeah, you can be forever young... if you're willing to travel at light speed.

"Now" Only Lasts 3 Seconds

The Ghost of Present Past

Brace yourself: your "present moment" is already gone. Psychologists suggest that our brains define "the present" as only about three seconds long. That's right, everything outside of those few seconds? It's either a memory or an anticipated future.

Think about it. When you clap your hands, the moment of clapping is already slipping into the past before your hands even finish meeting. Every word you're reading right now? Already history. The world around us isn't just fleeting—it's practically a ghost we're chasing in real-time.

And here's the kicker: if you meet a friend, the you that greeted them three seconds ago is already a memory. So are they talking to *you*, or just your ghostly, three-second-old echo?

Time Isn't What It Seems

The Elasticity of "Now"

Think you know what time is? Well, time might just be pulling your leg! Thanks to Einstein and the wonders of modern physics, we now understand that time is not a universal constant—it's more like a rubber band, stretching and contracting based on your frame of reference.

For example, if you're in a spaceship zooming through the cosmos at nearly the speed of light, your perception of time would slow down compared to someone chilling on Earth. In fact, at that speed, you could technically experience "now" in a way that's different from everyone else's—like getting a VIP pass to the universe's party!

But here's the kicker: even on Earth, our sense of time is all over the place. Ever notice how time flies when you're having fun and crawls during a boring meeting? That's not just your imagination; it's a real phenomenon influenced by your emotions and experiences.

So the next time you're stuck in traffic, remember: it's not that time is dragging—it's just being a bit flexible, giving you a chance to reflect on your life choices. What a generous cosmic joke!

Time as a Commodity

The Great Internet Time Heist

In our hyper-connected world, have you ever noticed how much time we spend online? Researchers estimate that the average person spends about seven years of their life scrolling through social media. That's a whole lifetime spent in the digital realm!

Keep in mind, time is treated like a currency online. Companies know how valuable your time is and often use sneaky tactics to keep you scrolling. Ever find yourself caught in a click hole? One minute turns into an hour before you know it, all because an algorithm decided to serve up a never-ending stream of cat videos and conspiracy theories.

So, what's the moral? While you might feel like you're just "wasting time," companies are actively banking on it. In a way, every minute you spend online is a little piece of your life being traded for likes, shares, and ads. It's a modern-day time heist, where your hours are the treasure!

Now that's a heist even Einstein would have a hard time wrapping his head around!

Before Time Zones
The Wild West of Clocks

Imagine a world where everyone kept their own time, and noon could mean different things depending on where you stood. Before the establishment of time zones in the late 19th century, that was reality! In fact, it was a chaotic patchwork of local time, with each town setting its clock based on the position of the sun.

In this "free-for-all" era, the clock tower in one town could chime noon while just down the road, another town celebrated lunch two minutes earlier. This meant travellers often had to ask, "What time is it?" at every stop, only to receive a bewildering array of answers.

But here's where it gets really wacky: trains! With the rise of railroads, scheduling became a nightmare. Trains would frequently collide or run late because the timetable relied on the wildly differing local times. It was like a cosmic game of telephone, but with more steam engines and fewer happy endings.

Finally, in 1884, Sir Sandford Fleming proposed the modern time zone system we use today, dividing the world into 24 zones, each one hour apart. This decision brought order to the chaos, allowing us to synchronize our clocks—and our schedules—across the globe.

So, the next time you check the time, remember: it wasn't always this easy! There was a time when even asking for the hour was an adventure—no app required!

Time and Navigation
The Longitude Lottery

Ah, the age of exploration—a time when the seas were vast, the maps were incomplete, and navigating was as much a gamble as a game of dice with fate. For centuries, sailors ventured into the unknown, struggling to pinpoint their longitude, the invisible lines that divide the Earth vertically. Without a way to measure time accurately at sea, they were like blindfolded adventurers, guided only by the stars, the tides, and a heavy dose of hope.

The stakes were high—countless ships were lost to shipwrecks, storms, or simply veering wildly off course. It wasn't just cargo and riches on the line; lives were at stake, and maritime empires desperately needed a solution. Enter the Longitude Act of 1714, the world's first innovation challenge on a grand scale. The British government dangled a tantalizing prize of £20,000 (think tens of millions in today's terms) for anyone who could crack the longitude conundrum. It was like the Olympics for inventors, but with fewer medals and way more drama.

Among the contenders was John Harrison, a humble clockmaker with big dreams and even bigger ideas. Harrison believed that the answer to navigation didn't lie in the stars or some complex mathematical equation, but in a precision timepiece—a marine chronometer. If sailors could carry a clock that kept accurate time, even through the rocking of a ship, they could compare local noon (when the sun was highest) to the time at a fixed point like Greenwich. That difference would reveal their longitude. Simple in theory, revolutionary in practice.

Harrison's obsession with precision led him to build a series of increasingly ingenious clocks, each one a marvel of craftsmanship and innovation. His designs faced skepticism, technical setbacks, and more than a little political intrigue, but his relentless pursuit of perfection changed the world. By the time his masterpiece, the H4, was completed, navigation had been transformed forever. No longer would sailors have to rely solely on celestial navigation or risk sailing blind—now they had time on their side, quite literally.

While Harrison was perfecting his chronometers, most sailors were left to rely on the age-old methods of using the stars, compasses, and gut instincts. It's humbling to think that they crossed vast oceans with tools that now seem quaint compared to the GPS in your pocket.

So the next time your smartphone reroutes you around traffic, take a moment to appreciate the brilliance of a clockmaker who made sure humanity could find its way—on land or at sea. And if you're ever lost in the middle of the ocean? Well, you'd better hope you packed a Harrison-grade chronometer...or maybe just download a good celestial navigation app before you set sail!

Ancient Sundials

When Time Was Literally "Made in the Shade"

Long before clocks, people relied on sundials to keep time, which meant your entire day depended on a giant stick casting a shadow. But here's the catch: sundials only work when the sun's out! So on cloudy days, time was anyone's guess—basically, "take a lunch break whenever you're hungry."

Ancient Romans even had public sundials, which sounds helpful until you realize that each one was set up based on local preferences. People would look at a sundial in one square and see "noon," only to walk down the road and find another sundial showing it was already 2 p.m. It made planning meetings or even keeping appointments practically impossible. And forget about sundials in the winter! People in northern Europe could lose entire *months* of sundial time.

Imagine telling your boss, "Sorry I'm late, the sun was hiding." It was the ultimate "blame it on the weather" excuse!

The Secret Lives of Body Clocks

Humans, Animals, and the "Time Zone Zoo"

Did you know that both people and animals run on built-in timekeepers? We have internal clocks that regulate sleep, hunger, and even mood—it's like having an invisible scheduler ticking away in your head. But the quirks of these clocks get downright hilarious in the animal kingdom.

Take the naked mole-rat: these strange, nearly blind rodents have no concept of day or night. Living underground, they operate on a "social clock." If one mole-rat decides it's mealtime, the rest just follow along, regardless of the hour. Imagine if your schedule depended on your hungriest friend!

Or look at pet cats—famous for their uncanny ability to wake you up at 5 a.m. sharp. Scientists believe cats have a special, finely-tuned "dawn clock" that tells them, "The humans are most vulnerable now... perfect time to demand food!" It's a skill they honed as wild hunters, but today, it just means your sleep doesn't stand a chance.

And humans? We might think we're modern, but our internal clocks are still wired for the outdoors. Ever feel like you're living in the wrong time zone? You're not alone! Our ancestors spent so much time around firelight that our bodies are convinced it's dusk as soon as we see candlelight. Light a candle tonight and watch yourself start to feel sleepy—proof that we're all just sophisticated animals at heart!

So, the next time you're up too early or too hungry, blame it on evolution. It's all part of the "Time Zone Zoo" we never quite escaped!

Time and the Divine

When Eternity Is Just the Beginning

Religion has always played with the concept of time, making it feel as elastic as taffy! In many faiths, the divine isn't bound by clocks or calendars, suggesting that the gods might view time the way we view a good book—something they can flip forward, back, or even pause for dramatic effect.

In some beliefs, eternity is as real as any ticking second. Think about it: if eternity is outside time, it's basically the ultimate "no-time zone" where even the most existential waiting room would be out of business! Imagine lining up at the pearly gates where there's no sense of time at all—you could arrive early, late, or "just in time," and it wouldn't make a bit of difference!

Then there's the idea of reincarnation, where you're recycled into life after life. It's like cosmic time-sharing—once you finish one lifetime, you just sign up for the next, whether it's a shot at ancient Egypt or a VIP pass to the distant future. If reincarnation is real, you're technically a time traveller on the universe's longest tour.

And who could forget the classic "end of days" prophecy? Many religions have versions of the end of time itself—a divine "closing time." But it's a curious paradox: if time ends, where does it go? Perhaps time is just taking a well-deserved break, relaxing somewhere in a dimension we can't see, sipping a cosmic cocktail and waiting to be summoned back.

Whatever your beliefs, one thing's certain: if there's a god out there with control over time, they probably have the most interesting watch collection in the universe!

Time Etiquette Across Cultures

The Art of Being Fashionably Late (or Early!)

Not everyone sees time the same way. In fact, cultural attitudes towards time vary so wildly that what's considered "on time" in one country might be hilariously off in another.

Take Germany, for example. Here, time is like an Olympic sport, with punctuality as the national pastime. Being five minutes late to a meeting? Practically scandalous. Germans see time as a well-oiled machine—everyone's expected to keep the gears turning. And if you show up too early? Don't! That's seen as interrupting their precious "preparation time."

Meanwhile, in Italy, time has a bit more... flexibility. Known as la dolce vita, Italians believe life (and time) is meant to be savoured, not scheduled down to the minute. Arriving late is practically expected, even polite—after all, no one wants to seem too eager. There's a joke that if you want an Italian party to start at 8 p.m., tell everyone it starts at 7!

Then there's India, where "Indian Standard Time" is often jokingly referred to as "Indian Stretchable Time." Here, time bends and stretches to fit social needs rather than strict schedules. If an event starts at 5 p.m., you might see guests trickling in at 6:30, and that's just fine. It's all about connection, not the clock.

Or consider Japan, where punctuality is almost a form of respect. Trains are so punctual that if one arrives 20 seconds late, it makes headlines. And if you're late for a social gathering? Apologies might be expected in person, by text, and by email!

So, if you're a "time traveller" bouncing between these cultures, remember: being "on time" means different things to different people. Just don't expect the same rules to apply everywhere, or you might end up lost in translation—or worse, the lone guest at an empty party!

Is Time Even Real?

Or Are We Just Really Good at Pretending?

Imagine for a moment that time isn't real. Yes, I know—sounds like the start of a sci-fi thriller. But plenty of thinkers, from ancient philosophers to modern physicists, have questioned whether time actually exists or if it's just a convenient illusion we use to make sense of things.

Let's get a little weird with it. In the grand scheme of the universe, "past" and "future" might just be our brain's way of organizing events. For all we know, all time could be happening at once, like an endless cosmic rerun. Right now, you might be reading this sentence at the same time the dinosaurs are roaming around, or at the same time you're sipping a futuristic coffee on Mars.

And what about the concept of "now"? It's so slippery that as soon as you think you've got it pinned down, it's already in the past. You can't hold "now" any more than you can catch a shadow. It's as if time is the ultimate prankster, constantly convincing us it's real while darting away just as we reach out to grab it.

One theory even suggests that we're all moving through something called "spacetime," where time is just the fourth dimension. So if you're feeling lost in time, don't worry—you might just be exploring a very confusing dimension. And the next time you find yourself running late, just tell everyone you're not late—they're simply perceiving time wrong.

So, does time exist? Maybe, maybe not. But until we figure it out, we'll just keep pretending it does, marking our calendars, setting our alarms, and trying our best to be "on time" for a universe that's laughing right along with us.

Timekeepers Through the Ages

From Burning Ropes to Atomic Clocks

Humans have tried measuring time with everything from shadow-casting stones to devices straight out of a steampunk novel. Forget your smartwatch; ancient civilizations had some seriously strange ways of tracking the hours.

Take the candle clock of medieval China. This brilliant (and slightly flammable) invention used candles marked with specific lines. As each line burned away, it signified a passing hour. Running late? Just burn a little faster! It's the perfect timekeeper—unless it's windy, or you happen to knock it over.

Or how about the clepsydra, aka the water clock, from ancient Egypt? The clepsydra was a pot that gradually filled or drained water to measure the hours. But it wasn't exactly reliable—water would evaporate faster in the summer, making the days feel shorter. And imagine hauling your clock with you on the go!

Fast-forward to the 20th century, when scientists took time measurement to the next level with atomic clocks. These ultra-precise marvels keep time by measuring the vibrations of atoms, usually caesium. They're so accurate that they lose only one second every 300 million years. Which means, if you ever want to be perfectly punctual for an appointment with a future civilization, the atomic clock's got your back.

But here's the twist: as techy as we've gotten, we still rely on natural events like the Earth's rotation to measure time. Ironically, we've discovered that even the Earth's "clock" is a bit wobbly—it's slowing down by about 1.4 milliseconds per century! So who knows? In the distant future, we might need to "re-calibrate" our entire concept of a day.

From burning ropes to vibrating atoms, it seems we've tried it all. And while our timekeepers have gotten more accurate, the basic truth remains: time always finds a way to keep us guessing!

Time Travel
The Myths, Legends, and the Bizarre Believers

Throughout history, people have been obsessed with the idea of time travel, creating myths, legends, and even some outlandish theories that have stood the test of time. Take, for example, the legend of Saint Augustine, who supposedly experienced a moment of divine "time dilation." According to lore, he walked into a garden in the morning and returned minutes later, only to find a day had passed—imagine trying to explain that to your boss!

Some of his ideas included:

Subjectivity of Time: Augustine proposed that time is not an external reality but a measure of the mind's experiences: memory (past), attention (present), and expectation (future).

Eternal Now: Augustine suggested that God's perspective transcends time entirely—God exists in an "eternal now," where past, present, and future are simultaneously present. In moments of divine connection, humans might briefly touch this perspective.

Moments of Divine Grace: Augustine viewed moments of divine grace or revelation as interruptions of linear time, offering a glimpse into eternal truths.

And then there are the urban legends. Perhaps you've heard about the time traveler who supposedly showed up with "modern knowledge" in the early 1900s, claiming to be from the future, complete with mysterious gadgets. Some believe in this man's story so fervently that they think he inspired many of today's "modern" inventions.

Or, if you're looking for scientific oddities, consider the "time slip" phenomenon that people claim to experience at famous landmarks like Versailles or in small villages in England. Visitors report accidentally "slipping" back in time, seeing people dressed in centuries-old garb and vanishing just as suddenly into the present. Scientists might shrug it off as imagination, but it sure makes you think twice about walking alone through an ancient castle!

With tales like these, time travel blurs the line between history, mystery, and myth. Even if we haven't cracked the code, these stories keep our dreams of time-bending alive.

Dreamtime

Where Minutes Become Hours and Logic Takes a Nap

Have you ever woken up from a dream that felt like it lasted hours, only to find you'd only been asleep for 10 minutes? In the strange world of sleep, time gets bent and twisted in ways that make the movie Inception look realistic.

In dreams, time is like silly putty—your brain stretches it, squeezes it, and even makes entire days pass in a single REM cycle. Ever had one of those dreams where you're late for a meeting, only to realize that you're still back in high school and can't find your homework? Somehow, your brain has compressed years of anxiety into a 30-second nap!

Scientists believe that this "time dilation" in dreams might happen because your brain's logic centre takes a break, giving your imagination free rein to mess with the clock. And it's not just in sleep—people who experience "flow states" (that intense focus when you're completely absorbed in a task) also report time feeling either faster or slower. It's as if your mind just decides, "We're rewriting the rules of time for a bit."

Dreamtime is like a parallel universe where time works on its own strange rhythm. So next time you wake up feeling like you've travelled through eons, just remember—it's all in a night's work for your brain!

Time Loops and Déjà Vu

When Life Feels Like a Rerun

Ever had that eerie feeling that you're reliving a moment you've already experienced? That's déjà vu, the uncanny sense that you've been somewhere or done something before—even though you know it's your first time. It's as if time is pulling a prank on you, making you think you're stuck in a mysterious loop.

Some scientists say déjà vu happens because of a little "misfire" in the brain, where new experiences get briefly mixed up with old memories. Others suggest it might be a result of your brain processing something too quickly, causing you to "remember" it before you've fully experienced it. In a sense, you're tricking yourself into feeling like you're in a time loop. Thanks, brain!

There are legends, though, that see déjà vu as something spookier: a glitch in the "matrix" of reality, or even a hint that we're living in a simulation where past events get repeated now and then. And if you ask science fiction fans, they'll say déjà vu is just the universe's way of giving us a sneak peek at parallel timelines.

So, is déjà vu just a funny brain quirk, or a sign that we're all secretly in a time loop? Either way, the next time you get that weird "I've been here before" feeling, just remember—it's all part of time's never-ending sense of humour!

Chasing Immortality

The Timeless Quest to Outsmart Aging

Humans have always been obsessed with staying young, and history is full of wild ideas for halting time. From mystical fountains to bizarre beauty treatments, our ancestors tried every trick in the book—and it was often a very weird book.

Take the legend of the Fountain of Youth. People in the 16th century were so obsessed with this mythical spring that explorer Ponce de León supposedly scoured the Americas searching for it. Spoiler alert: he didn't find it. But the idea that there's a magical cure for aging is as persistent as time itself.

In ancient China, emperors downed elixirs of immortality that included, among other things, mercury (not the best anti-aging tactic in hindsight). It was thought that drinking these mystical potions would grant eternal youth, but instead, many ended up aging themselves to the afterlife a bit faster than planned.

And let's not forget the Victorians, who were big on beauty concoctions. They invented all sorts of "anti-aging" creams made from ingredients like crushed pearls and silver dust.

Did they stop time? Probably not. But at least they had a radiant look, even if it was mostly from arsenic powder.

These days, with science backing us up, people still try to "pause" aging, from cryogenic freezing to anti-wrinkle lasers. Yet time keeps winning, no matter what we throw at it. Maybe the lesson here is to embrace a little laugh line or two—they're just markers on your personal time map, after all!

Talking Time When Languages Bend the Clock

Time might be universal, but how we talk about it is anything but! Different languages view time in ways that would make your head spin—and some cultures even see time in shapes we'd never expect.

Take the Aymara people of South America. For them, the future isn't "ahead"—it's actually behind them! They reason that we can "see" the past because it's already happened, so it's in front of us, while the unseen future is creeping up from behind. Just imagine explaining that to your boss next time they ask about your "forward-thinking" vision. This isn't just a poetic metaphor. Their language reflects this unique relationship with time. In Aymara, words for "past" often align with terms for "in front," while "future" aligns with "behind." It's a spatial map of time that feels so alien yet makes perfect sense when you think about it: We can "see" the past because it's already played out before our eyes. The future, however, is hidden, like an unwelcome guest sneaking up from behind.

Imagine applying this concept perhaps in the realm of sports, it adds a twist to the advice,

"Keep your eye on the ball!"—maybe they should tell you to watch your back instead.

This reversal of time's flow challenges us to rethink the assumptions we take for granted. If time isn't a straight line—or even a forward march—what else might we be getting wrong about the nature of reality? The Aymara's perspective may seem unusual to us, but they might argue that our obsession with "moving forward" is just as peculiar.

If you thought time was always a straight arrow or a winding river, Mandarin Chinese offers a surprising twist. In this language, time flows vertically, like water cascading down a mountain. Instead of moving forward or backward, the past and future are mapped to "up" and "down," creating a unique spatial perspective.

For example, in Mandarin:

- The word 上 (*shàng*), meaning "up," is used to describe the past. 上个星期 (*shàng gè xīngqī*) means "last week," and it feels as if you're looking up at

something that has already passed, like the summit of a mountain.
- Meanwhile, 下 (*xià*), meaning "down," refers to the future. 下个月 (*xià gè yuè*) means "next month," suggesting that the future lies below, waiting to unfold as you descend into it.

This vertical view of time offers a fresh perspective on life. The past isn't "behind us" but above us—something we've already climbed and can now look back on with clarity. The future, on the other hand, is below us, a slope yet to be traversed, full of uncertainty and possibility.

The roots of this concept might be found in Chinese philosophy. Time in Mandarin reflects the cyclical nature of life, where yin and yang and the Taoist flow govern everything. Each moment isn't just a tick on the clock—it's part of a grand stack of layers, each one building on the other.

Imagine explaining this concept to your boss during a brainstorming session: "Where do you see yourself in the next quarter?"

"Well," you might say, "I see myself further downhill, hopefully gliding rather than tumbling!"

Jokes aside, Mandarin's vertical mapping of time feels grounded, literally and figuratively. It encourages respect for the layers of history above us while reminding us to tread thoughtfully into the unknown depths below.

Next time you hear someone say, "It's all downhill from here," you might think of Mandarin Chinese and smile. After all, downhill could simply mean you're heading toward a future full of momentum and opportunity.

In the English-speaking world, time is on the move—always charging ahead, with the past behind us and the future stretched out before us like an open road. This forward-backward orientation of time is so deeply embedded in the language that we rarely stop to question it.

Consider common phrases like:

- "Looking ahead to the future"
- "Leaving the past behind"

- "A step forward in time"

These expressions reinforce the idea of time as a journey along a straight path, where the past is something we've already passed and the future is something we're approaching.

This perspective is even reflected in the way we speak about personal growth or progress. Being "forward-thinking" is a compliment, while being "stuck in the past" is decidedly not. Success is often framed as "moving forward" in one's career or life, and setbacks are described as "steps backward."

English also has a competitive streak when it comes to time. Phrases like "running out of time," "beating the clock," or "saving time" suggest that time is a resource to be managed or even conquered. It's a race, and we're all expected to keep up.

But this relentless forward march can create pressure, too. Time in English doesn't just flow—it surges. There's rarely a moment to pause and reflect because, as the saying goes, "time waits for no one."

While the forward-backward view of time may dominate, English speakers also use metaphors that hint at other ways of thinking. Time can "fly," "drag," or even "stand still," depending on our emotional state. In these moments, we're reminded that our experience of time isn't always tied to the clock—it's something deeply personal and fluid.

This perspective raises an intriguing question: Is our linear view of time a reflection of reality, or is it just a product of how our language shapes our thinking? Could time be less of a journey and more of a dance, a loop, or even a still point?

For English speakers, time is a relentless march forward, ticking off the seconds like a metronome. But for the Hopi people of North America, time isn't something you measure—it's something you experience. The Hopi language doesn't even have a word for "time" as we understand it. Instead, their view of time is deeply tied to events and their unfolding, a fluid perspective that challenges everything we think we know about the passage of time.

In the Hopi language, the focus isn't on past, present, or future but on whether something is manifest (visible and knowable) or unmanifest (hidden and not yet knowable). For example:

- Events that have already happened or are happening now are treated as manifest. These are tangible, like the rain falling from the sky or the crops already planted in the ground.
- The unmanifest includes things that haven't yet occurred, like tomorrow's weather or next year's harvest—possibilities that exist in potential but haven't yet come to fruition.

This doesn't mean the Hopi people don't recognize the passage of time; they simply prioritize the reality of events over the abstract ticking of a clock. Their language reflects a worldview rooted in cycles of nature, patterns of life, and the rhythms of the Earth.

Imagine living in a world where the concept of "being late" doesn't exist because time isn't measured in minutes or hours but in actions and readiness. The Hopi approach to time invites us to think less about the numbers on a clock and more about the natural flow of life. It's a perspective that feels profoundly freeing—and perhaps a little intimidating.

Next time you find yourself stressed about a looming deadline, you might consider the Hopi view. Instead of asking, "How much time do I have left?" ask, "Is this the right moment for this to happen?" It's a subtle shift, but one that could change your relationship with time entirely.

Whether you see time as a road ahead, a stack of layers, or even a shapeless flow, the way we talk about it shapes the way we experience it. In the end, time is just one more thing that's lost in translation—and that might be the best way to understand it!

Comyn/31 Things to Know About Time

Memory
The Time Machine
in Your Mind

Our memories might be the closest thing we have to a personal time machine. But unlike the sleek machines of sci-fi, memory is more like a wobbly, glitchy, emotionally charged ride that doesn't always go where we'd like. Scientists have discovered that our brains don't just "record" events like a camera; they actively recreate them each time we recall them. So every time you reminisce about that holiday five years ago, you're effectively warping it with new layers, blending in details that might not even be true.

Memory is one of the biggest culprits in time's distortion. Not only does it ignore the idea of a fixed timeline, but it also plays favourites. Enter the reminiscence bump—a quirky psychological phenomenon where our 20s and 30s shine in ultra-high definition, while other decades feel like they were recorded with a shaky hand and poor lighting. Researchers think this bump exists because those years are packed with firsts—first job, first love, first mortgage-induced panic attack—all forming a highlight reel in our minds.

What's fascinating is how these years set the emotional tone for much of our lives. The music you obsessed over during this time? It's probably still the soundtrack of your inner world. The movies? Timeless classics, even if they were critically panned. And the friendships? They feel etched in stone, even if some have faded into the abyss of Facebook memories.

The bump isn't just about nostalgia, though. It reflects the brain's way of prioritizing events that shape our identity. Those years of exploration, triumphs, and awkward missteps create a mental map that helps us navigate who we are. It's like the brain decided, "Alright, these are the golden years. Let's burn these into the archive and let the rest simmer in the background."

And yet, the reminiscence bump exposes memory's creative license. While it feels like we're replaying the past, the truth is, we're just stitching together fragments. Every time you recall a moment, your brain edits it slightly—adding details, omitting others, and sometimes outright fabricating parts. So the vividness of your 20s and 30s? It's not just a highlight reel; it's a *remix*.

In the end, the bump serves as a reminder that memory isn't a perfect historian—it's a biased filmmaker. It tells the story it wants you to believe, complete with the greatest hits from your prime-time years.

Ah, false memories—the brain's most mischievous party trick. These aren't just minor mix-ups, like forgetting where you left your keys. No, false memories are full-blown fabrications, vivid and detailed, of events that never happened. Scientists have shown just how malleable our memory is, proving that with enough suggestion and imagination, you can be convinced of things that are entirely fictional.

How does this happen? It turns out your brain isn't a faithful recorder—it's more like a creative storyteller. Every time you recall a memory, it's less like opening a file and more like rewriting a scene. Each retelling comes with a chance for embellishments, omissions, or outright mistakes. Sprinkle in some external influence—

say, a convincing story from a friend or a well-placed suggestion—and voilà! A brand-new memory is born.

Psychologists have even created false memories in the lab. In one famous study, researchers convinced participants they had once been lost in a shopping mall as a child. How? By weaving the false story into real events from their lives and encouraging them to imagine it. Many participants eventually recalled vivid details of this fictional event, swearing it had actually happened.

It gets weirder. Repeated exposure to a story—whether through family anecdotes, photoshopped images, or even your own daydreams—can convince your brain to adopt it as truth. Picture yourself vividly on a childhood vacation to Paris, walking along the Seine, eating a croissant. Do this often enough, and your brain might decide it's a genuine memory, even if you've never been to France.

But false memories aren't all bad—they're just a by-product of the brain's remarkable flexibility. This creative tinkering is what allows us to imagine the future, empathize with others, and even tell great stories. It's as if your brain is less concerned with accuracy and more interested in creating a narrative that feels cohesive.

So, the next time someone insists they saw you wearing a lime-green tuxedo at a wedding you know you skipped, remember: they might not be lying. Their brain just made a mistake in the editing room.

Nostalgia is more than just a fond memory—it's an emotional teleportation device. A familiar song, the scent of an old perfume, or even the taste of a childhood treat can press "play" on your brain's time-travel soundtrack, whisking you back to moments you thought were long buried. It's not just recalling the past; it's re-experiencing it, often with the emotional volume turned up to maximum.

When nostalgia strikes, it's like your senses open a hidden door to the past. That summer road trip with friends comes rushing back with the crackle of an old playlist. The smell of freshly baked cookies? Suddenly, you're standing in your grandmother's kitchen, watching her pull a tray from the oven. These sensory triggers bypass logic and take a direct route to the heart, making nostalgia feel less like a memory and more like a reunion with your former self.

What's fascinating is how nostalgia doesn't just bring back the "what" of an experience—it also rekindles the "how." It resurrects the joy, excitement, or bitter sweetness you felt at the time. It's why hearing a song from your teenage years doesn't just remind you of that school dance; it makes you feel 16 again, awkwardly shuffling in your best (or worst) outfit.

Scientists believe nostalgia is more than just a pleasant distraction. It's a psychological tool that helps anchor us in our identity. In moments of uncertainty or stress, it reminds us who we are and where we've been. Studies even suggest that indulging in nostalgia can boost mood, increase resilience, and make us feel more connected to others. It's a little like a mental comfort blanket, warming us with memories of times when life felt simpler—or at least, rosier in hindsight.

And that's the thing about nostalgia: it's not always historically accurate. The brain has a habit of editing the past, smoothing over the rough edges and polishing up the highlights. Those carefree summer days? Probably not as carefree as you remember. But nostalgia isn't about accuracy; it's about meaning. It's less concerned with the truth of the past and more focused on how it makes you feel now.

So, when nostalgia sneaks up on you, let it. Let that song, smell, or taste take you on a journey. It's your brain's way of giving you a break from the present, offering a brief escape into a past that feels as vivid and alive as the day you lived it.

In the end, memory may be an imperfect, hazy time machine, but it's the one we carry with us wherever we go. And if it occasionally rewrites history a bit, maybe that's just its way of keeping life interesting!

Time and Rhythm

When Music Messes with the Clock

Music is built on time, but it has an almost magical ability to play around with it, bending and shaping it in ways that make a few minutes feel like a lifetime or an instant. Ever noticed how a fast-paced song can make a long workout fly by, or how a slow, sombre tune seems to expand time? Musicians have been toying with time for centuries, and they've developed all kinds of tricks to keep us on our toes.

Time signatures are the hidden architecture of music, the patterns that dictate how a song flows through time. Most of the music we hear marches along in familiar rhythms like 4/4 or 3/4, the comforting structures that feel natural to our brains. These rhythms are so ingrained in us that they often mirror our own bodily rhythms—walking, breathing, or even a heartbeat. They give music a predictable flow, a steady pulse that anchors us in the moment.

But what happens when music steps off the beaten path? Enter the world of odd time signatures like 5/4, 7/8, or even more complex patterns. These unconventional rhythms break the mould, shaking up our expectations and forcing our minds to adapt. Instead of the even, predictable beats of 4/4, these irregular time signatures introduce a sense of unpredictability, as if the music itself is playing with time.

Take 5/4, for instance, where each measure gets an extra beat. The result feels like the music is leaning forward, never quite settling into the groove. A classic example? Pink Floyd's "Money"—its choppy, off-kilter rhythm perfectly complements the song's themes of greed and materialism, making the listener feel slightly unbalanced, just as the subject matter intends.

Or consider 7/8, a rhythm that feels almost like it's tripping over itself. It's as if the music is perpetually catching its breath, creating a sense of tension and urgency. Bands like Radiohead thrive in these unusual rhythmic landscapes, using them to evoke feelings of unease, wonder, or even cosmic disconnection. Tracks like "2 + 2 = 5" or "Pyramid Song" transport listeners into a space where time bends and warps, mirroring the emotional complexity of the lyrics.

Odd time signatures aren't just technical curiosities—they're emotional tools. They force listeners out of autopilot, making us pay closer attention. When a song doesn't conform to the rhythms we expect, we're drawn into its world, compelled to navigate the unfamiliar terrain. It's a musical way of saying, "Hey, pay attention—time isn't as simple as you think."

Tempo is one of the most powerful tools in a composer's kit, the speed at which music moves that can dramatically alter how we experience time. It's like the engine of the song, propelling us forward or slowing us down, creating the illusion of time stretching or contracting, depending on the pace. Ever notice how a slow ballad can make moments feel heavier, like time itself is crawling, while an upbeat track can make the minutes fly by? That's tempo working its magic.

In slower songs, like ballads or mournful orchestral pieces, the slower tempo mimics our natural tendency to dwell on thoughts and emotions, making us more aware of each passing moment. It's as though time is stretching out in sympathy with the song's reflective mood. Take a classic like Eric Clapton's "Tears in Heaven"—its sombre pace amplifies the emotional weight, making each note feel deliberate, as if time itself is measured in heartbeats.

On the flip side, fast tempos inject a sense of urgency and excitement. They compress time, making everything feel quicker and more intense. Think of dance tracks, high-energy rock anthems, or fast-paced classical movements. The rapid beats don't just get our bodies moving; they also trick our minds into perceiving the passage of time in a condensed, more exhilarating way. When a song speeds up, it feels like we're running alongside it, in sync with the beat—time suddenly feels like it's sprinting, pushing us forward.

But tempo doesn't just follow a simple pattern—it can shift. The genius of composers and bands is how they manipulate tempo to create emotional landscapes and control the listener's experience of time. One of the most iconic examples of this is Queen's "Bohemian Rhapsody". This six-minute song takes you on a sonic rollercoaster, using multiple tempo changes and shifts in time signature to craft a journey that feels longer, or sometimes shorter, than its actual length. The ballad section moves slowly, with operatic swells, creating a dreamlike feel where time seems to stand still. Then, the rock section bursts in with a sudden, fast tempo change, shifting the energy and speeding up the passage of time. The transitions between these varied tempos and rhythms take us through a labyrinth of emotions and experiences, leaving us breathless by the end, all within the span of just a few minutes.

This manipulation of tempo is not just for effect—it mirrors the way we experience time in our daily lives. Sometimes, life seems to drag when we're waiting for something or in a quiet moment of reflection. Other times, it speeds up when we're caught up in excitement, running through the chaos of daily tasks, or simply living in the present. Just like a tempo shift, these moments stretch and compress based on our emotional state.

Ultimately, tempo is more than just speed. It's about creating a rhythm that resonates with our experience of time, helping us feel, understand, and connect with the passage of moments, whether fast or slow. And when it changes unexpectedly, it can make us feel like we're living in a whole new timeline altogether.

Music's power over time goes beyond just tempo and rhythm; it taps directly into memory and emotion. Certain songs are so closely tied to specific moments in our lives that they can whisk us right back in time, making a three-minute tune feel like a portal to the past. It's as if every beat and note holds a sliver of our own personal timeline, waiting to bring it back in an instant.

In the end, music doesn't just measure time; it shapes it, stretches it, and folds it over. Whether through tempo, rhythm, or nostalgic pull, every song is a reminder that time is as much about feel as it is about numbers on a clock.

Comyn/31 Things to Know About Time

Biological Clocks

Nature's Strange Timetables

Time isn't just a human concept; nature has its own way of keeping track, and it turns out animals experience time in ways that seem straight out of a sci-fi novel.

Take the Greenland shark, for instance. This slow-moving creature, drifting through the cold waters of the North Atlantic, can live for over 400 years! Imagine being born in Shakespeare's time and still kicking around to see the invention of smartphones. Greenland sharks live life at a pace so slow that time almost seems to stand still for them.

Then there's the opposite extreme: the mayfly, whose adult life can last as little as 24 hours. In a single day, it hatches, matures, finds a mate, and passes on—all in what seems like the blink of an eye. For a mayfly, a minute is precious, each second a tiny eternity.

Meanwhile, animals like cats and dogs seem to have a time-sensing superpower. Their circadian rhythms are so well-tuned that they often know when it's time for dinner down to the minute. Some researchers believe that animals may even have a better sense of "short time" than we do—like measuring seconds and minutes, which explains why your dog is waiting at the door the second you're usually home.

But perhaps the most fascinating timekeepers in the animal world are migrating birds and butterflies. Each year, the monarch butterfly travels up to 3,000 miles from Canada to Mexico, somehow sensing the exact time and route to get there, even though no single butterfly completes the round trip. It's as if they have a "time map" imprinted in their genes.

For animals, time is woven into the rhythms of life and survival. They may not check clocks, but from sharks to mayflies, every creature lives by a unique biological timetable. In the end, time is relative—even if you're just a dog waiting impatiently for dinner.

The Speed of Light

When Time Stands Still (Sort Of)

Time and the speed of light have a relationship that feels like science fiction, yet it's very real. The faster you move, the slower time ticks—at least according to Einstein's theory of relativity. If you could travel at the speed of light, time would theoretically stand still.

Imagine hopping aboard a spaceship and zooming off at light speed (about 299,792 kilometres per second). From the outside, people would see you frozen in time. Meanwhile, you'd feel normal, but when you returned, everyone else would have aged while you stayed exactly the same. It's a cosmic loophole that could let you skip ahead in time—if only we had the tech to travel that fast.

Light itself is a bit of a time traveller, too. The light from distant stars we see today has often taken millions, if not billions, of years to reach us. So when we look at the night sky, we're really seeing deep into the past. In fact, if someone on a planet 65 million light-years away had a super telescope and looked at Earth, they'd be watching dinosaurs roam.

The speed of light is the fastest thing we know of, and it sets the ultimate limit for how quickly information and energy can travel. But what does this have to do with the ancient pyramids? At first glance, they may seem unrelated, but there are a few fascinating connections that arise when you dive into the mysteries of both the speed of light and the pyramids of Egypt.

One of the more mind-blowing ideas comes from the precise measurements of the Great Pyramid of Giza, the only surviving wonder of the Seven Wonders of the Ancient World. Some researchers, particularly those intrigued by the connection between ancient knowledge and modern science, have suggested that the pyramid's dimensions reflect an awareness of astronomical measurements, including those of light and distance.

For example, the latitude of the Great Pyramid, at approximately 29.9792458° North, is eerily close to the speed of light in kilometres per second (299,792.458 km/s). This uncanny coincidence has sparked speculation about whether the ancient Egyptians had some knowledge of the speed of light, thousands of years before scientists like Einstein brought it into our modern understanding. Was it just a coincidence? Or could the pyramid's architects have encoded these measurements into the structure as a kind of cosmic blueprint?

While there's no direct evidence that the Egyptians had knowledge of the speed of light in the way we understand it today, the alignment of the pyramid's dimensions with this number seems remarkably precise for something built over 4,000 years ago. Could the Egyptians have had access to some form of ancient wisdom or astronomical insight that we've yet to fully understand? Or perhaps it's all just a coincidental alignment of numbers that happen to match?

Beyond the numbers, the pyramids also have a symbolic connection to light in a broader, spiritual sense. In ancient Egyptian culture, the sun was central to their religious and cosmological beliefs. The pyramids were not just monumental tombs; they were symbolic gateways to the afterlife, often aligned with the rising and setting of the sun, which was seen as a divine force of creation and renewal. The pharaohs, upon their death, were believed to join the sun god Ra in the afterlife, and the pyramid itself was thought to act as a vehicle to propel the soul toward the heavens.

The way the pyramid is constructed—its shape, its orientation, and its perfect alignment with cardinal directions—suggests that its creators were intimately aware of cosmic phenomena, and possibly of the light of the stars, the sun, and the moon. They were, in a sense, building a connection between earthly existence and the celestial realms, potentially even symbolizing the speed with which a soul might travel to the heavens—whether through light, energy, or divine force.

From a more speculative standpoint, some enthusiasts have linked the idea of the speed of light to concepts of time and space, particularly in the context of how ancient structures may have been built with astronomical precision. The alignment of the pyramids with the stars—especially the Orion constellation, which was closely associated with the Egyptian god Osiris—has led to the idea that the pyramids could be a kind of "cosmic timepiece," capturing the essence of universal forces like light and time itself.

Could the Egyptians, in some way, have encoded the relationship between light and time into their monumental structures? This remains a mystery, but there is a compelling argument that the pyramids were designed to embody something far beyond the physical realm, hinting at universal truths about space, time, and light.

While we can't say definitively that the ancient Egyptians knew about the speed of light in the scientific sense we understand today, the alignment of the Great Pyramid's latitude with the speed of light is one of those strange and mystical coincidences that continue to provoke thought. Whether it's the geometric precision of the pyramids, their connection to the cosmos, or their symbolic role in the afterlife, the pyramids remain a testament to the ancient quest to understand time, space, and the light that governs the universe.

Then there's gravitational time dilation, which shows that time moves more slowly in stronger gravitational fields. This means time ticks slightly slower at sea level than it does on top of a mountain—though the difference is so tiny you wouldn't notice. But scale it up to the immense gravitational pull near a black hole, and time could slow down almost infinitely. So if you ever find yourself orbiting a black hole, just remember—your friends on Earth might have aged quite a bit when you return!

In the end, light and gravity bend time in ways that seem almost magical. While we might not be able to travel at light speed or visit a black hole anytime soon, these cosmic quirks remind us that time is as stretchy, bendy, and mysterious as the universe itself.

Aging in Time

The Strange Clock Inside Us

Ever noticed how time seems to fly by faster the older you get? As a child, each year feels like an eternity, but as we age, those same years seem to slip away in the blink of an eye. This isn't just a perception issue—there's real science behind it. When you're a kid, a single year is a huge chunk of your life. At two years old, one year is half of your entire existence! But by the time you're in your 30s, 40s, or beyond, a year is just a tiny fraction of your accumulated life experiences. This shift is partly why time feels like it's speeding up as we grow older.

But the passage of time isn't just a mental trick. It's deeply embedded in our biology, in ways most of us don't even realize. Our bodies have their own internal clocks that tick away as we age, and surprisingly, some of these clocks are very precise—like the ticking of a metronome, but one that eventually runs out. You might have heard of these internal timekeepers: they're called telomeres. Telomeres are the protective caps at the ends of our chromosomes, and with every cell division, they shorten ever so slightly. As they get shorter, our cells start to show signs of aging. This process is thought to play a big role in why we age. It's like every cell in our body is carrying a tiny hourglass, slowly running out of sand.

In a sense, we're all living with our own internal countdown clocks, ticking away with each cell division. It's a phenomenon that has fascinated scientists for decades, and has led to some ground-breaking discoveries. For instance, research has shown that by manipulating these telomeres, it might be possible to slow down the aging process or even extend lifespan. While we're still a long way from reversing aging in humans, some animals have shown increased lifespans when scientists tweak their telomeres or other genetic factors. So, although the idea of "turning back the clock" may seem far-fetched, it's not entirely out of the realm of possibility.

But here's the real twist: the process of aging isn't uniform across your body. If we could zoom in on the cellular level, we'd find that different tissues age at different speeds. For example, brain cells tend to age much more slowly compared to other cells, while skin cells, which undergo frequent turnover, have a much faster renewal cycle. This means that, in a way, time is ticking at different rates in different parts of your body! Your skin might feel the full force of aging, while your brain is still ticking along relatively untouched.

In a sense, we're all walking around with a body full of "clocks," each one ticking at its own rhythm. The rate at which we age may seem to be speeding up as we get older, but it's actually a complex interplay of biology, perception, and the way our cells and tissues experience time. What we experience as "time flying by" is merely a reflection of our shifting perspective on time itself—and the fascinating biological mechanisms beneath that experience.

Lost in Space (and Underwater) How Extreme Environments Warp Time

In extreme environments, like the vast reaches of space or the dark depths of the ocean, time can feel...different. Take astronauts on the International Space Station, for example: they orbit Earth every 90 minutes, experiencing 16 sunrises and sunsets each day! This rapid-fire day-night cycle means that "day" and "night" blur together, often causing astronauts to lose track of time. To combat this, they follow strict schedules and even use special lights that mimic the Earth's day-night rhythm to help keep their internal clocks in sync.

Underwater, deep-sea divers report similar time distortion. In the depths, sunlight barely penetrates, creating a timeless, twilight zone where it's hard to tell day from night. Divers often emerge thinking only a few minutes have passed, only to find they've been submerged for hours. The lack of normal daylight cues seems to slow their sense of time, creating an eerie feeling of timelessness.

These environments highlight just how much our bodies depend on familiar signals like light to keep time. Without them, our internal clocks start ticking to their own, unpredictable rhythms. Even the simple act of telling "how long" you've been somewhere becomes a puzzle.

In space, the effects go further. Due to gravitational time dilation, astronauts in orbit around Earth age just a tiny bit slower than people on the ground. It's not much, but over a lifetime, a few months in space would actually make them milliseconds younger than their Earth-bound friends—time travel at a very, very small scale.

Extreme environments remind us that time isn't only about clocks and calendars; it's about how our bodies perceive it. And whether underwater, in space, or here on Earth, our experience of time can stretch, shrink, and transform in ways we'd never expect.

The "Butterfly Effect" and Time

How Small Actions Can Change Everything

You've probably heard the phrase, "A butterfly flaps its wings in Brazil, and it causes a tornado in Texas." This whimsical image captures the heart of the Butterfly Effect, a fascinating concept from chaos theory that suggests small, seemingly insignificant actions can trigger massive, unpredictable consequences. It's a vivid reminder that our lives are intricately woven into the fabric of time, where even the smallest decisions might send ripples through the future.

Imagine this: one morning, running late, you decide to take a different route to work. That simple change—just a few extra minutes in your day—could set off a chain of events. Maybe you miss a bus, bump into an old friend, or avoid an accident you would've otherwise been involved in. A single choice, seemingly inconsequential, alters your path and the course of your day, if not your life.

The Butterfly Effect challenges our traditional view of time as a linear, predictable sequence. Instead, it suggests that time is more like a vast, intricate tapestry, with each decision or event acting as a thread that interweaves into a complex, interconnected web. Chaos theory teaches us that even the tiniest of actions can accumulate and create entirely new timelines—meaning that a small choice, while appearing insignificant in the moment, could be the catalyst for monumental changes down the road.

When we bring time travel into the mix, the Butterfly Effect introduces an enticing paradox. Picture a time traveller who goes back to stop a seemingly trivial event—perhaps a friend's minor car breakdown. By altering this one small detail, they could inadvertently trigger a chain reaction that distorts the future in ways they couldn't have anticipated. The result? Catastrophes, unexpected shifts in history, or even the erasure of their own existence. Films like Back to the Future and Predestination have explored this concept, highlighting the mind-bending consequences of meddling with time.

So, while we often view time as a straight line, the Butterfly Effect encourages us to imagine it as a swirling cloud of infinite possibilities. Every choice, no matter how small, has the potential to ripple through the fabric of time, creating a cascade of potential futures.

The Psychology of Time

Why Does Time Fly When You're Having Fun?

Have you ever noticed how time flies when you're having fun but drags on during a boring lecture? This curious phenomenon is rooted in our psychology, and it reveals just how subjective our experience of time can be.

When we're engaged in enjoyable activities, our brains release a cocktail of neurotransmitters like dopamine, creating a feeling of happiness and excitement. This heightened state of engagement leads us to focus deeply on the present moment, causing time to feel like it's speeding up. In contrast, when we're stuck in situations that are dull or stressful, our brains may hyper-focus on the clock, making seconds feel like hours.

This "time dilation" effect can even influence our memories. Events that are packed with thrilling experiences often feel longer in hindsight. Think about an amazing vacation filled with new adventures—it might feel like you were away for weeks, even if it was only a few days. Our brains tend to remember moments that are rich in emotion, novelty, or excitement more vividly, leading to the illusion that more time has passed.

Interestingly, researchers have found that age can play a role in our perception of time as well. Children tend to experience time more slowly because, for them, every year is a significant portion of their lives. As we grow older, each year becomes a smaller fraction, making time feel like it's speeding up.

And here's where it gets even weirder: our perception of time can actually be manipulated. In experiments, scientists have shown that simply changing the colour of a room or the rhythm of music can alter how fast or slow participants feel time is passing. It's as if we're living in a reality where the mind holds the ultimate power over time!

So, the next time you find yourself in a tedious meeting or a thrilling adventure, remember that your perception of time is as much about your brain's tricks as it is about the actual passage of seconds. Time is not just a ticking clock; it's a canvas painted by our emotions and experiences.

The Time of Your Life
How Moments Make Memories

We all talk about "the time of our lives"—those moments that become the highlights of our memories. But what makes certain experiences feel like they're worth a lifetime while others fade away? It turns out, the way we remember time is different from how we live it.

This is thanks to something psychologists call the "Peak-End Rule." It's a brain hack where we tend to remember experiences based on their most intense moments and how they ended, rather than the entire experience. That's why a rough road trip that ends with a spectacular sunset might still be remembered fondly!

Another reason we have certain "times of our lives" is nostalgia, which brings its own strange time-warping effect. When we recall our "glory days," we're often piecing together moments, adding a bit of shine, and sometimes even tweaking details to make them feel extra special. Studies show that nostalgia not only makes us feel warm and fuzzy; it also has the strange ability to make those memories seem as if they happened "just yesterday," even if it's been decades.

Then there's the feeling of time speeding up with age. The phenomenon called "The Reminiscence Bump" explains why we remember events from our teens and twenties so vividly; it's a time of "firsts"—first loves, first jobs, and first big adventures. Our brains are more likely to tag these new experiences as significant, which makes that phase of life feel rich and full in our memories.

In the end, the "time of your life" isn't really a single stretch of time—it's a blend of peak moments, nostalgia, and all those cherished "firsts" that come together to create a personal highlight reel. And maybe that's the real magic of time: it allows us to relive the best parts over and over, at least in memory.

Time and Magical Thinking

From Lucky Charms to Time-Turning Spells

Humans have always been captivated by the idea of controlling time. We're the only species that doesn't just live in it—we're constantly trying to hack it, extend it, and sometimes even outsmart it. Throughout history, magical thinking has influenced how we see time, from ancient rituals to modern "lucky charms" that people swear will keep them on time or give them a longer life.

Consider the rituals of ancient civilizations that believed they could influence the seasons, like summoning rain or sunlight. Many early cultures would perform time-altering ceremonies to encourage longer days for crops to grow, convinced they could make time itself cooperate. In medieval Europe, people sought the Philosopher's Stone—an alchemical substance that could supposedly grant eternal youth. It's the idea that time doesn't have to be a relentless force; it can be tamed, softened, or even rewound.

And who hasn't wished for a magic spell to give them a few extra hours? The idea is timeless:

even today, people knock on wood to "pause" bad luck or carry a lucky item to avoid a mishap. It's as if, deep down, we believe a simple object or gesture can momentarily pause or reshape the ticking clock.

In modern fantasy stories, time manipulation takes on new life with enchanted items like time-turners or wishing wells. These devices resonate with us because they speak to our deepest wishes: to relive a favourite moment, undo a mistake, or simply have more time. Even science fiction plays on this with time machines, hinting that, just maybe, technology could give us a kind of magical power over time.

Whether it's a charm, an amulet, or just crossing our fingers, magical thinking around time shows how much we long to interact with it in mystical ways. It's a reminder that, while we can't control time, we'll always look for little ways to feel like we can. Because isn't life a little more magical with a dash of possibility?

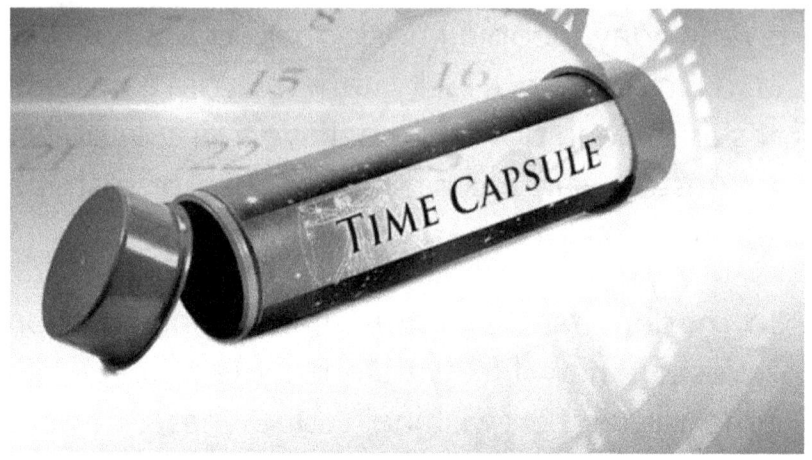

Time Capsules

Messages to the Future

For centuries, people have been leaving messages in bottles, boxes, and even in the walls of buildings, hoping that someone in the future would uncover them. These are time capsules—a fascinating way for past generations to reach forward through time and share a slice of their world with us.

Time capsules come in many forms, from personal notes and photos buried in backyards to elaborate community efforts, like the Crypt of Civilization in Georgia, USA, which was sealed in 1940 and contains everything from classic novels to beer. Designed to stay closed until 8113 AD, the Crypt is a "mini museum" of the early 20th century, packed with items selected to help future humans understand our way of life.

There's a strange hopefulness in time capsules; they're often filled with objects and notes meant to tell a story of everyday life, as if the creators are saying, "We were here, and this is what mattered to us." They include mundane items—newspapers, everyday tools, even snacks—because they capture daily life better than grand monuments or history books.

Some capsules, however, reflect very particular moments in time. Take the Westinghouse Time Capsules of the 1939 and 1964 New York World's Fairs, which include things like microfilm newsreels, a pack of cigarettes, and letters from prominent thinkers. The creators intended these items to convey the worries, hopes, and accomplishments of their era. The capsules are set to be opened in 5,000 years, long after most traces of our world will have vanished.

Of course, not all time capsules are found when they're supposed to be. Sometimes they're discovered too early or even accidentally lost forever. In 2017, a school time capsule from 1987, filled with toys, a video tape, and student letters, was discovered purely by chance during renovations—30 years too early! These chance discoveries give us a little peek into the past, often with items that feel both familiar and oddly nostalgic.

In a world that's always moving forward, time capsules are a rare, tangible link to those who've gone before us, allowing them to connect across centuries. Maybe they remind us that while we live in different times, there's a common thread in our shared humanity—and maybe, that's the most timeless message of all.

Time in the Movies

Hollywood's Take on Time Travel and Beyond

The silver screen has brought us countless tales that defy the rules of time, bending it, looping it, and sending characters into alternate pasts and futures. Movies let us imagine what it would be like to jump backward, forward, or to live the same day over and over, exploring both the thrill and the consequences of manipulating time.

One of the most iconic examples is Back to the Future. Its DeLorean time machine made us dream about visiting the past—and trying not to mess it up. Marty McFly's adventures showed us that even small actions can have big consequences. It popularized the "butterfly effect" in time travel, where stepping on a bug in the past could change the future in unpredictable ways.

Another classic, Groundhog Day, plays with time in a different way: the time loop. Poor Phil is stuck living February 2nd over and over until he learns some life lessons (and that a time loop can actually drive you a bit mad). Time loops have become their own genre, giving us a peek at what it would be like to perfect every detail of a single day, or to be forced to make a fresh start until we finally get things right.

Then there's Interstellar, which takes time to the cosmic level, exploring relativity and the strange, heart-wrenching effects of time dilation. When Cooper travels to a planet where an hour equals seven years back on Earth, we're hit with the emotional toll that time can take when it's stretched by the laws of physics. Interstellar reminds us that time isn't just something we live by—it's something that can profoundly alter relationships and lives.

Modern sci-fi movies like Inception and Tenet push the limits of alternate time dimensions, playing with dreams within dreams and the idea of experiencing events in reverse. They challenge our minds and force us to confront the possibility that time may not be as straightforward as it seems. With layers upon layers of past, present, and future, these films leave us wondering: is time something we can control, or are we at its mercy?

Movies let us break free from time's usual boundaries and imagine a world where we could alter, slow down, or even stop it. Whether it's by rewinding like Doctor Strange, jumping to the past like Avengers: Endgame, or experiencing endless tomorrows like Edge of Tomorrow, these films allow us to escape the constraints of time—even if just for a few hours in the theatre.

In a way, movies about time aren't just about the past or future; they're reflections of our own hopes, fears, and fascinations with the unknown. They let us explore the "what ifs" without ever needing a time machine, asking questions about what it would mean to actually control our own time. And maybe that's the magic of movies: they create their own little pocket of time, letting us experience lives, loves, and adventures that are timeless.

THE END

www.ingramcontent.com/pod-product-compliance
Lightning Source LLC
Chambersburg PA
CBHW071034240526
45469CB00006BD/2204